What This World Is Really About!

What This World Is Really About!

Because the Confusion Starts with the Fundamentals

By

Ken Minton

Strategic Book Publishing and Rights Co.

Strategic Book Publishing and Rights Co.
12620 FM 1960, Suite A4-507
Houston TX 77065
www.sbpra.com

For information about special discounts for bulk purchases, please contact Strategic Book Publishing and Rights Co. Special Sales, at bookorder@sbpra.net.

ISBN: 978-1-63135-476-2

Acknowledgements

To my wife, Leslie, for all she does, including assistance with this endeavor.

To Mr. Peter N. Schrup, for his gifts of "scientific" publications.

To Dr. Donald Urquidi, for his knowledgeable input into chapter one.

To Mr. P. J. Williams, for his positive review of the preliminary chapters.

To Dr. Sid Nirenberg, for his encouragement regarding chapter one.

To Deanna and the Edit Team, for all their efforts.

Contents

Preface

Before anyone assumes things, let me explain why I wrote this book, over a twenty-year period, with the audacious title of *What This World Is Really About!* To start with, I was basically forced into the college system by my sweet, loving mother. Let me make this clear, it was not something I had always wanted to do. As a result, I had no particular direction. Thus, I studied philosophy, took psychology, and reviewed political science. I acquired a minor in architecture. Eventually, I received a degree in mechanical engineering from San Diego State University and advanced forward on my required path to work.

But, as I proceeded through life, I thought to myself, *Was college just vocational?* Clearly, the system buried me in a tremendous amount of data. But, did it teach me or any other student how to understand reality? If so, why is there is so much disagreement in the world? The institution does present philosophy and science. But does this breakdown actually make sense? To me, the answer was an emphatic no. Thus, I proceeded on my twenty-year journey, which resulted in this book.

Let me make this clear, with this book I challenge Western thinking at the very rudimentary core. Chapter one is an analysis of those fundamentals. With an agreement on how reality is understood, chapters two through seven address many of this world's controversial issues, the very issues the academics have failed to conclude. Chapter eight is a review of their perspectives.

A chronological list of prominent Western "philosophers" is presented. Finally, chapter nine is included as a summary.

In turn, if the critics challenge my analysis, let them have at it. Some academics have already tried and failed (See chapter one for their comments). Thus, I am quite confident any reader will be hard pressed to find inconsistencies, as I found in the academic's philosophy versus science breakdown. That said, I present – *What This World Is Really About!*

Chapter One

Because the Confusion Starts with the Fundamentals

As you, the reader, may have noticed, this chapter reads, "Because the Confusion starts with the Fundamentals." But what confusion you ask? This author argues, the ubiquitous confusion that permeates modern societies because Western academic thinking is flawed at the fundamental level. But how is it flawed? If the question is posed, how is reality understood? Whether realized or not, the academic's answer is delegated to either the scientific method or the philosophic method whereby the scientific method is supposedly testable (or predictable) while the philosophic method is not. Therefore, certain issues are scientific, and thus provable, while the remaining issues are consequently, philosophic, and thus unprovable. Unfortunately, this thinking is incorrect!

Problem one: In regards to test ability or predictability, there often exists no actual differences between some stated "scientific" issues and some stated "philosophic" issues. Let us consider a specific scientific issue. As an example, the big bang theory, is presented as scientific. But, are scientists testing the entire theory by creating a new universe? Of course, the answer is no. Modern technology is able to test only aspects of the theory. Consider the scientific theory of evolution. Is life being created at the cellular level with the necessary allotment of time, billions of years, which allows evolution to create another human species? Clearly, the answer is no. Again, only some aspects of the theory are testable or predicable.

Now, let us consider a classic philosophic issue. Does human free will exist? Or is determinism correct? Are there any potential tests? The answer is there are numerous potential tests. For example, drop a stone here on planet earth, what will happen? The stone falls to earth. Whether realized or not, this physics-related test is an indicator to the possibility of free will. Please know every time the stone drops as predicted cause and effect is substantiated. If cause and effect is 100 percent, then what is assumed to be free will is a myth. In other words, all human choices in a complete cause and effect universe are predetermined. Consequently, there is no "room" for free will. To the contrary, quantum mechanics, which governs atomic physics, seems to support probability. And, testing is possible in this realm as well. Consequently, this so-called philosophic issue will be concluded by scientists, not philosophers. So why is this presented as a philosophic issue?

Similarly, the philosophers have something they call a priori knowledge. This is knowledge before and independent of sensory experience. In order to verify this as true, scientific testing was conducted on very young, unlearned children. But the scientific testing verifies the demarcation between philosophy and science makes no sense.

Now let us consider a "scientific" theory. Please consider Einstein's famous general relativity theory explaining gravity. Please know that general relativity theory was published back in May of 1916 after a 1913 published Einstein-Grossmann paper. In fact, it was not until 1919 that any confirmation of the theory was substantiated with astronomical observations of light from the sun, which were found to be curved. All additional details of the theory remained fully untested until 1970 when hydrogen maser clocks confirmed spacetime warping as a result of earth's mass. Consequently, was Einstein's theory just philosophy for fifty years? Please, know said testing did not confirm all aspects of general relativity. It was not until 2007 that NASA's Gravity Probe B made further testing possible. In fact, Einstein's space-time continuum theory suggests space in the presence of mass creates a geodetic effect, which causes gravity, as finally tested by NASA. But, secondly, his published theory predicted a rotating mass drags along space-time causing "frame dragging." This aspect of the famous theory remains untested. Finally, please know general relativity was first published as a theory, not a hypothesis.

Now consider the nine (so far) theories regarding parallel universes. Please know these theories propagated in order to explain the known facts of our universe. In Brian Greene's book, *The Hidden Reality*, on page nine he writes, "The subject of parallel universes is highly speculative. No experiment or observation has established that any version of the idea is realized in nature" (Green, p. 9). Consequently, is the issue of parallel universes philosophy or science? Additional evidence of this confusion is also on page nine where he makes the reader aware that one of the theories has – "origination in the philosophical community." So, is that particular theory "philosophic" and the others are "scientific"? Clearly, that conclusion would make no sense. Please understand these presented examples (and there are lots more) are contradictory to the demarcation separating philosophy from science.

Problem two: A currently untestable issue may become testable with future technology as evidenced with Einstein's general relativity theory.

Problem three: Additionally, the semantics of philosophy and science are different. This leads to confusion as if two different methods can be acceptable for understanding reality. Different methods are likely to lead to different conclusions, so this makes no sense.

Problem four: The academics' demarcation fails to separate out human like and dislike from what is correctly true about reality. Their solution – empirical versus a normative value proposition – will be presented later in this chapter.

Problem five: All evidence supports reality as an interconnected unified system. Consequently, the demarcation of philosophy versus science erroneously severs a unified system. This makes no sense.

Problem six: When there is ample related evidence a test is unnecessary to confirm a theory. For instance, if a conventional gasoline powered motor car has no fuel, will the vehicle function? Clearly, the answer is no. Therefore, a test is not always required in order to verify a theory. Consequently, would this specific issue be that of philosophy or science? Either way it would not make sense.

So, what does make sense? When analyzed, there exist three very different types or categories of issues. There are issues about reality, issues about preference, and issues about fiction. Because no controversy exists in regards to fiction, the remaining two categories of issues are the subject of this book. But what is a reality issue versus a preference issue?

A reality issue is the way reality, this all-inclusive environment, actually was, is, or can be. For example, the numbers four plus one equals five. The numbers do not equal six or seven or anything else. Likewise, a wood table is made from a tree or other plant substance. It is not made from stone, brick, or other non-plant substances. Thus, for a given reality issue, there corresponds a correct answer consistent with reality; therefore, all other answers are incorrect or inconsistent with reality. Other than for subjectivity, the academics confirm this with their grading of student's answers as correct or incorrect.

To the contrary, preference issues are about what we, emotional living beings, like or dislike about reality or our perception of reality. All choices and all values, those of morality and ethics, as well as music and art, fall under this category. Unlike the reality issues, no correct or incorrect answer exists with the preference issues (see chapter five for the caveat regarding right versus wrong and good versus bad). For example, some people may like or prefer the taste of cow's milk while other people may not. There is no correct answer here. Though, some may argue milk is beneficial to an individual's health, a reality issue, nevertheless, an individual may dislike the flavor, a preference issue. Without this analysis or break down of information, confusion is a certainty.

But there is additional confusion in regards to the preference issues. Often preference issues are expressed as absolute. As examples, "that girl is beautiful" or "that girl is ugly." Both are actually incorrect statements. You may like the girl's looks or you may not like the girl's looks, but there is no absolute to this or any other preference issue. That is, the girl is not actually beautiful or ugly, as those are individual preference descriptions. In contradiction, the PhD philosophy department argues "scientific" testing has verified certain absolutes to beauty because very young, uncultured, or unlearned children, unanimously responded negatively to unsymmetrical body shapes. But their thinking is faulty. Whether the programming for like versus dislike is in our genetics or in our culture, does not alter the issue from being that of preference. For instance, an alien life form could be programmed differently from us where they prefer unsymmetrical body shapes.

Other examples of preference issues are as follows: Are taxes too high? Are taxes too low? Who is the best? Who is the worst?

Should guns be legal or illegal? Should abortion be legal or illegal? Should drugs be legal or illegal? Should there be a death penalty? What is the meaning to life? These are just some examples of preference issues. And the preference issues have no correct or incorrect answer. But that conclusion doesn't make these issues insignificant. Clearly, morality and ethics play a vital role in society. Nevertheless, intellectually there is no absolute correct answer in regards to preference issues. That said, inconsistencies do occur when hypocrites attempt to promote preferences that affect others that they themselves will not accept.

With the two types of issues established, we must next establish the method for an accurate assessment. That is, how do we distinguish between what is consistent with reality, thus true, and what is not? The French philosopher Descartes attempted to get absolute knowledge certainty with his statement, "I think, therefore I am." But how did he know with absolute certainty he was not in a dream at that very moment? If so, was he actually thinking or just dreaming that he was thinking? Thus, this author would argue (as others have done) absolute certainty is a myth. Nevertheless, we can make sense of our exposure, limited by our senses, to the reality information.

Because no one sees totality, only pieces, the method for understanding is essentially dictated. That is, reality is a puzzle with pieces of information we must put together. There is no other logical method. But how do we know any single piece is true? We know a single piece is true only by its fit with the other pieces. If you doubt me, consider the word "true" is only significant if something else is "untrue." Furthermore, all words are defined in terms of other words. Thus, a comparison is required. It can't be avoided. For instance, how do we know four plus three equals seven? With the previous definitions of the quantities of three, four, and seven, as well as the meaning of addition and equals established, we then count the total of three plus four and compare it to the quantity of seven. And of course these two quantities are, as fact, equal, but it is *not* by definition; it is by comparison. Next, we build the reality puzzle by intellectually placing the directly related true pieces or facts together.

What is a fact? A fact is a piece of information presented as true that will hold up to scrutiny. Furthermore, that piece of information can encompass raw data or previously processed information. Let it be understood facts include the following: evidence, statistical data, physical phenomenon, and other data such as measured quantities. But facts are not limited too such. Though no fact is known with 100 percent confidence, it is nevertheless our most certain type of information; nevertheless, let it be clearly understood that any fact can be challenged. But if we are in agreement upon the related facts, we proceed with a theory. Then, where possible, we test the theory in order to create more test facts that may or may not agree with the original theory. But, tests are not necessarily a pass or fail as many would falsely assume. The actual value of a test is that more related facts are produced.

What is a theory? A theory binds the pieces or known facts together. That is, theories are *explanations*, which account for the directly related facts. Please consider a common definition: (1). A coherent group of general propositions used as principles of *explanation* for a class of phenomena; (2) A proposed *explanation* whose status is still conjectural and subject to experimentation. Please note the word "phenomena" refers to the related facts. Also note, a theory is an explanation as opposed to a guess. And also please note, some theories are more speculative in that they explain fewer facts than others. Arranged in logical order the progression would be as follows: guess (few or no facts), speculative theory (some facts), viable theory (many facts), and factual explanation (accounting for all known directly related facts where no reasonable alternative theory exists). Please know all predictions fall into one of these categories.

Importantly, in many circumstances more than one theory may explain the same set of facts. This would be typical for speculative theories where few facts are available. However, where more facts are known competitive theories often fail leaving a winning theory by the process of elimination. Please know in some circumstances a factual gap may exist. Additionally, many complex issues entail sub-theories to the main theory. For instance, if an individual is a suspect to a crime by theory, one sub-theory might explain his means of entrance while another sub-theory might explain his means of escape.

In regards to theories, there are essentially two types of theories, a specific-case theory, and a general-case theory. As an example, when released from elevation, will a specific glass structure break at impact? For this, a test would prove it one way or the other. However, a general-case theory is different in regards to proof. For example, when released, will all stones everywhere drop to the ground? A test in this case would only substantiate, not prove, the theory because there could be untested stones somewhere else that may refute this general-case theory. Thus, for general-case theories, testing results do not prove, testing results only potentially disprove. Science is predicated upon this fact. Thus, the scientific method does not prove; the scientific method disproves. There is general confusion in this regard. Nevertheless, this author states that when a general-case theory has significant substantiation, and there exists no reasonable competing theory, then the theory itself must be moved into the factual explanation column. Otherwise, building the reality picture would be limited. Please know even scientists do this without admitting so.

Furthermore, many general-case theories can be expressed in measurable terms. That is, an equation is but an algebraic representation of a general-case theory. And a factor, a component of an equation, is the general-case version of a fact. For example, your height might be as fact, five feet, eight inches, but height in general is a factor. Thus, the two words, fact and factor are integral to each other. Additionally, most equations or general-case theories have a dependent factor, the outcome, contingent upon more than one factor. What this means in everyday life is when outcomes are being considered, it is necessary to change only one factor at a time. In other words, intelligent thinking is, what is the outcome when all other factors are equal?

Finally, let it be clearly understood there is discovery, but no invention. For instance, math is not a construct of human invention as many educated people would falsely assume. What we call the number of toes at the ends of our feet is a result of counting a specific quantity. We are not inventing the number of toes; to the contrary, we are merely naming that specific quantity. Likewise, algebra and calculus are the *measurements* of relationships as well as cause and effect in the "real" world. In fact, calculus was discovered by two different individuals, Newton and Leibniz, at roughly the same time frame because algebra had been previously discovered. If Newton and Leibniz had never lived, someone else would have made the calculus discovery. Likewise, did the Wright brothers invent the airplane, or did they discover that wings cause lift when speeding through air molecules? In conclusion, what humankind calls invention is actually complex discovery.

Now that I have presented the two issues, reality issues and preference issues, as well as the basic method for assessment, how does this compare to the academic's perspective? In similarity to reality issues versus preference issues, academia presents objective versus subjective. But an objective issue versus a subjective issue is not a correct substitute for a reality issue versus a preference issue. In the modern world, the term objective actually refers to the non-bias state of the observer, not the observed issue. Thus, it would make no sense to say "that table or that chair is objective." Consequently, no issue is actually being proclaimed or acknowledged because the word objective refers to the observer, not the observed. Therefore, the word "objective" issue suggests the observer must remain objective, which is not the same as stating the type of issue being addressed.

Furthermore, the word "subjective" is not a correct substitution for the word "preference." This is because the term "subjective" refers to both inaccurate perception and preference. For example, if a person sees the color green, a specific frequency of light, as yellow, this would be correctly referred to as his or her subjective perception. This would not be his preference to see green as yellow! Thus, there is a clear difference between having inaccurate perception and having a preference or bias. But, the word subjective includes both meanings.

Likewise, consider two swimming pools, one and two. When measured with a thermometer pool one is at forty-two degrees while pool two is at eighty-six degrees. Now, ask a person to feel if the pools are cold or hot. Based upon the person's subjective feelings he says, "Pool one is colder than pool two." Because cold versus hot is relative, but are also measurable terms, this issue is a reality issue with a correct answer. That is, forty-two degrees is colder than eighty-six degrees. On the other hand, asking that person which pool he prefers to swim in would be a preference issue with no correct or incorrect answer. Furthermore, if the two pools were kept at close but different temperatures, say forty-two degrees and forty-four degrees, the person using his subjective feelings may incorrectly reverse cold and hot. Thus, his subjective perception is less accurate than a precision measuring device, but in this case, is not a preference. Consequently, the terms objective and subjective do not correctly account for the existence of reality and preference issues.

After years of debate, hidden away in the esoteric halls of PhD philosophy, there is a presentation of empirical proposition versus a normative-value proposition (their words) where their word empirical might be substituted for reality issue and normative-value might be substituted for preference issue. To make this clear, PhD philosophers have seen the obvious need (like this author) for separating out value (subcategory to preference) from statements of factual truth about reality. Therefore, from their standpoint, an empirical statement is that which can be determined to be true or false. Whereas normative-value refers to "subjective" values and interpretations. However, a problem occurs if they combine values (preferences) with subjective interpretations. The length of a stick could be used as an example. Without an accurate measuring tool, a person might "guess" the length, using his subjective interpretation capabilities. If the philosophers were to put this under the normal-value category, they would contradict their own test of true or false because the subjective guess could be compared to an accurate measurement and thus proven true or false. The conclusion is preferences must be separated out from the other issues or contradictions are inevitable. The academics acknowledge this fact, but erroneously seem to fuse inaccurate perception with issues of preference.

Finally, let it be understood a philosophy acknowledging the existence of reality has been previously put forward. Nevertheless, if the explicit acknowledgment of a preference issue relative to a reality issue is absent, the big picture is flawed.

In regards to the educational system's explanation of how knowledge is gained, how reality is understood, as previously stated, there is the problem with the answer of philosophy versus science. Clearly, scientists study only reality issues, but leading to confusion is the fact that philosopher's study both reality and preference issues. In order to understand their demarcation, please know science developed from natural philosophy, which in turn, developed from philosophy. And it was the German-educated philosopher, Immanuel Kant (1724-1804 AD) who suggested that certain issues should be designated under traditional philosophy and separated from natural philosophy (predecessor to modern science). Unfortunately, he got the demarcations wrong! To make this clear, Immanuel Kant was educated in the same faulty thinking that has plagued every Western philosopher from day one. Consequently, instead of correctly separating out preference from reality, he incorrectly severed an interconnected reality system into seemingly testable issues versus seemingly untestable issues. Thus, certain reality issues are currently studied by philosophers and other reality issues by scientists. But proof of the mistake is evidenced in the fact certain reality issues currently labeled under philosophy actually have significant scientific input as previously mentioned with the issues of determinism and a priori.

Additionally, modern intellectual philosophy as presented by the educational system is divided into four branches. That would be metaphysics, axiology, logic, and epistemology (study of knowledge). Intellectual metaphysics, different from astrology-associated metaphysics, is known as the first philosophy. Essentially, this branch of philosophy is associated with Rene Descartes' pursuit of knowledge certainty while searching for reality's essence. David Hume, the prominent English philosopher (1711-1776 AD), realized the obvious – absolute certainty is not possible. Nevertheless, the problematic word "opinion" was designated as different from real knowledge. Apparently only they, institution members, generate theory while we nonmembers mindlessly produce mere opinions. What makes this a problem leading to confusion is "opinion" is also used when expressing preference. For instance, it is Bill's opinion the core of the moon is made of iron, a reality issue, and it is Bill's opinion he likes moon light at night, a preference issue. Clearly, the former is a theory about reality while the latter is a stated preference. What would make sense is the word theory should be designated to reality issues and the word opinion to preference issues. If neither category of issues were specified, the word "perspective" would make sense, as it is Bill's perspective.

In regards to an axiology, it is the study of morality and ethics. These are issues. However, in the broader scope, these two are but subcategories to the larger issue of human values. But this author will argue human values are, in turn, a subset to the larger issue of preference. Those in academia are failing to use inductive reasoning to extend from the specific case to the most general case. Furthermore, morality as stated by the institution is "relative." Though correct, morality is actually relative to someone's preference.

As far as logic and epistemology, this author would argue both are related to the issue of reality. That is, if there are other realities, the logic used here may not apply there. For instance, if A equals B and B equals C, than we take A to equal C. But it is possible, in another unforeseen reality that may not be true. Consequently, logic is tied to the comparison of information we take to be true or consistent with reality. Thus, our exposure to reality is dictating the methods of logic. This in turn, leads to epistemology. But what are the methods of philosophy and science that result from epistemology?

The methods of philosophy are as follows: There is an intellectual argument based upon stated premises with the conclusion accounting for the premises whereby the definition of a premise is an assertion. If you substituted the word fact for premise and the word theory for conclusion, the philosophic method would be similar to what this author presented previously. However, there is nothing in the philosophic method specifically suggesting all available directly related facts must be addressed. This is a problem for the philosophic method. Please see chapter eight of this book for more inconsistences with philosophic thinking.

So what are the methods of science? According to Dr. Tim Berra, a professor of zoology and a former editor of the Ohio Journal of Science:

> The scientific method involves the observation of phenomena or events in the real world, the statement of a problem, some reflection and deduction on the observed facts and their possible causes and effects, the formation of a hypothesis, the testing of the hypothesis (experimentation or prediction), and – when the tests repeatedly confirm the hypothesis the erection of a theory. (Berra, p. 2)

This is printed on page two of his book, *Evolution and the Myth of Creationism.*

If we analyze Dr. Berra's explanation of the scientific method, it is fundamentally in agreement with this authors' explanation of how reality is understood. Please note, Dr. Berra implies the observations of phenomena are the observed facts. Then a hypothesis is formulated. Let it be clear, the hypothesis is the theory; it is just the initial theory. Importantly, testing may or may not alter the hypothesis. Please know there are numerous theories that stand identical to the original hypothesis. Additionally, in science, when a new theory replaces an existing theory it is called a paradigm shift.

In regards to science in the medical field, clinical studies are performed in order to identify contributing factors. A control group is separated out in such a study in order to determine a correlation. In turn, a high correlation suggests evidence of a contributing factor. Nevertheless, because other factors are always at play, these correlations cannot be taken as proof. Again, there is no proof in science.

But there are two significant problems with the *understanding* of science, which causes confusion. Firstly, scientists don't spend millions of dollars testing a hypothesis, the initial theory, if it fails to explain all of the related facts in the first place. That is, such a theory is already a *failure* without the need for testing! Furthermore, please understand if only test facts are to be considered, a divorce from reality is eminent. Though test facts are important, a potentially correct theory must explain *all* related facts. This thinking is reinforced by the very definition of a theory as previously presented.

The second problem with the *understanding* of science is as follows: At no point does science, in terms of semantics, proceed from theory to factual explanation. That is, the word "theory" is the verbal end point. There exists no verbal progression to "factual explanation." In his book Dr. Berra explains on page four, "A scientific theory is the endpoint of the scientific method." Unfortunately, most people are unaware of this. Consequently, many would incorrectly argue, What about Newton's laws of motion? Newton's laws of motion are actually the observed facts. Furthermore, all thermodynamic laws are the observed facts. But what is called theory in science never proceeds to factual explanation. Again, science only disproves; it does not prove. Unfortunately, even science teachers can be unaware of this.

Finally, in regards to science, inherent is the requirement of a test potentially disproving a presented theory. For example, the theory that states when a stone is released it will fall to the planet has the potential of being disproved. That is, when released, the stone could remain stationary or move upward or sideways. Now let us consider a theory that states whatever should happen is caused by "the man in the moon." In this case no potential failure is possible. Whatever happens, as stated by the theory, is caused by "the man in the moon." Clearly, any such theory is problematic, but this author emphasizes the lack of related facts substantiating a "man in the moon" automatically fails this problematic theory. Unfortunately, there exists an equally problematic theory in major acceptance by humankind. That would be, the Almighty, Creator, 'God' theory. Like it or not, the existence of an *assumed* God is a theory about reality with our understanding dependent upon related facts. But why is it being sold as a "faith" issue? The explanation will be presented in chapter five.

In conclusion, what might potential critics say to my presentation of reality versus preference issues? Specifically, if the critic were to suggest my analysis is just one of many subjective approaches, as one professor did, to that I would answer, is it a subjective approach to acknowledge the existence of the sun? Furthermore, is it a subjective approach to acknowledge the existence of the moon? And finally, is it a subjective approach to acknowledge the difference between the sun and the moon? Of course the answer is no to all of the above. Likewise, reality versus preference, as issues, cannot intelligently be denied.

But why would the academics not acknowledge this? Please know it is not coincidental that the institution's demarcation of philosophy versus science happens to provide them with job security. That is, please save the hard thinking for the *paid* professional scientists and philosophers.

If I have ruffled some feathers, so be it. So far, numerous professors have attempted to pick apart my presentation, and while failing to do so, they have exposed their own confusion. But in the process, they have exposed the very need for such a book. As far as my presentation of only two issues (other than fiction) is concerned, I am merely taking one more step than the library. The library categorized or separated fiction from nonfiction. Please imagine the confusion if they didn't. I am merely extending this analysis. Thus, the preference issues must be separated from the reality issues. Furthermore, the PhD philosophy department arrived at a similar conclusion, the need to separate value, but their semantics, empirical versus normative-value, is alien and problematic. And for job security, that department will never admit there are no "philosophic" issues any more than the Pope will admit there is no God or the Professional Wrestling Association will admit their wrestling is fake!

Clearly, this author sees much confusion in the modern world. The problem starts with the failure of the educational system to get the fundamentals correct. Let me make this clear. There are no scientific issues; there are no philosophic issues! Rather, there are issues about reality and there are issues about preference. In fact, people called "scientists" study certain reality issues. Please know it would make more sense to call them, "researchers." Furthermore, it would make more sense to call science, "research." Nevertheless, while adding to the confusion, people called "philosophers" study both reality issues and preference issues. Consequently, when someone says, "that's an unknowable philosophic issue," what do they mean? Are they referring to a preference issue or are they referring to what they think is an untestable reality issue? But, if their reference is to the latter, their thinking is misconceived. Nevertheless, when the average person uses the word philosophy he really means preference. For instance, when someone states, "my philosophy is–live and let live," this is actually a statement of preference. Clearly, academia has left the world's population confused by promoting scientific versus philosophic issues instead of correctly promoting reality versus preference issues.

So, if there is confusion with these fundamentals, what are the ramifications? Read on.

Chapter Two

The Fact of Evolution

In review of chapter one, you, the reader, should know the following: there are no issues of philosophy or science. Correctly, those are, in fact, methods, not issues. To the contrary, the actual issues are those about reality and those about preference. Furthermore, you now know any valid method of assessment requires an explanation (a theory) to account for all of the related facts, including those produced by testing when available. But any explanation that fails to account for the available related facts is faulty without the need of testing. Please remember, scientists don't waste millions of dollars testing a hypothesis, the initial theory, if it fails to explain all of the related facts in the first place. With this in mind, let us consider the evolution debate, an issue about reality, not preference.

Historically, this controversial reality issue has been argued with certainty by the proponents of evolution because it's scientific. But a counter argument opposing evolution has been put forth by an academic suggesting Darwin's theory is nothing more than unknowable philosophy. But is any of this true?

In regards to how the human species came to be there are only a few possible explanations. This fact does not elude everyone. In his book, *Darwin on Trial,* author Phillip F. Johnson, a Berkeley law professor and a self-proclaimed Christian, attempts to create doubt about Darwin's explanation in regards to biological evolution. By the process of elimination the obvious implication is, the only acceptable alternative is that an intelligent "God" creation must be true.

So, what are the only possibilities to explain how you, I, and all other life came to exist on this planet? The possibilities are, there was a naturally caused evolution over billions (thousands of millions) of years, or there was a purposeful intelligent creation as the religious would have us believe, or life on this planet was started by intelligent aliens from a distance location. It needs to be noted life could have started here or elsewhere initially.

From chapter one of this book, a theory is an explanation accounting for the related facts, which encompasses the related evidence. To his credit, Phillip Johnson does present many of the related facts associated with the issue. This would include the fossil record, embryology, genetics, and anthropology. But the failure of his presented logic was to assert his own competing detailed explanation or theory, which accounts for the facts he himself presented. Throughout his book, he alludes to "Gods" creation; consequently, let us consider this explanation first in regards to the stated facts. Please, consider in the Christian sense a God is the Almighty as well as the Intelligent Designer. If he exists, he is the Creator. But do the related facts Phillip Johnson concedes support the God theory, a theory about reality?

On page fifty-four of *Darwin on Trial*, Johnson presents the following:

> In recent years evidence of bacteria and algae has been found in some of the earth's oldest rocks, and it is generally accepted today that these single-celled forms of life may have first appeared as long as four billion years. Bacteria and algae are "prokaryote," which means each

creature consists of a single cell without a nucleus and related organelles. More complex "eukaryote" cells (with a nucleus) appeared later, and then dozens of independent groups of multicellular animals appeared. (Johnson, p. 54)

These are his words. They are the conceded facts.

Please note, the simplest forms of life, single-celled without a nucleus, first appeared about four billion years ago, then more complex (in his own words) cells appeared later. Actually, the oldest fossils found to date are of Cyanobacterial (prokaryote), which formed dolomite in the Pilbara range in North West Australia. According to the BBC, these very tiny (an electron microscope was required for observation) organisms were dated at about 3.5 billion years old. According to Wikipedia, the free encyclopedia, the first complex (with a nucleus) single-celled life formed about 1.8 billion years ago. In LiveScience, posted February 04, 2009, by the senior editor, Robin Lloyd, the earliest multicellular organisms might have been as far back as 1.2 billion years ago. Thus, it required one point seven billion years before the first jump from a simple organism without a nucleus to one with a nucleus. The second step in complexity to multicellular took an additional six hundred million years. Now, consider the Almighty, Creator, God theory.

If God exists and is the all-powerful creator, why not produce all life forms at once rather than leave the appearance of a step-by-step evolution from simplest and smallest to slightly more complex, which required billions of years? Could it be the "natural" cause and effect required the significant expanse of time in order for environmental conditions to adequately support the more complex creature? If so, why would a Creator be required? That is, if a supposed Creator is constrained by the laws of physics which apparently required more than one thousand million years for more complex creatures to appear, then the Creator is not the Almighty. If the supposed Creator is not the Almighty then how could he create something from nothing as the religious would have us believe? Thus, the Almighty, Creator, God theory is in dire conflict with the conceded facts.

Additionally, on page forty-nine, Phillip Johnson states the following:

> The discovery of Archaeopteryx–an ancient bird with some strikingly reptilian features—was enough fossil confirmation in itself to satisfy many. Thereafter, it was one apparent fossil success after another, with reports of human ancestors, ancient mammal-like reptiles, a good sequence in the horse line, and so on. (Johnson, p. 49)

Let it be clear at no place in his book does he explain away these conceded facts. So what is his basic argument?

In Darwin's time, an adequate knowledge base was already in place supporting the fact a lengthy evolution was involved in the shaping of life. It was Darwin who suggested the natural mechanism in which the evolution could have occurred. That mechanism would be "decent with modification," as he called it or "natural selection," as we know it. Though Phillip Johnson accepts on page sixteen of his book that natural selection "has an effect in maintaining the genetic fitness of a population," (Johnson, p. 16) he rejects natural selection can produce new species. He argues by the standards of scientific test ability that Darwin's natural selection has failed to produce new species. But let us consider what is meant here by test ability. In many circumstances a test can be conducted in a relatively short time frame. For instance, testing the effects of gravity can be performed in quick order because gravity's effects are fast acting. In the case of evolution millions of years were required in order to produce new species. Thus, evolution was very slow acting. Consequently, at this point in time no new species have been produced in the laboratory. But, when geneticists do artificially create a new species in the future, Johnson already rejects the results because it will have been done artificially using human intelligence. Thus, testing of the theory for the creation of new species is not a possibility by his standards. He has effectively put the theory in a no-win situation. Otherwise, we will potentially need to wait millions of years to verify a new species develops naturally. In the mean time, his book sells and makes him money while pleasing numerous Christians.

But, is Darwin's theory unprovable "philosophy" as Johnson explicitly claims in his book when full testing is effectively unavailable? If so, then Einstein's general relativity theory and the big bang theory are likewise just philosophy. Let it be understood that Johnson's argument about a lack of test ability could easily apply to most "scientific" theories. Thus, Johnson inadvertently exposed the faulty demarcation of philosophy versus science. And remember, science produces theories, not factual (as in proven) explanations, anyway. Clearly, everyone is confused. This brings us back to chapter one of this book. Darwin's "natural selection" is a theory about a reality issue, not a preference issue, where a theory is defined as an explanation accounting for the related facts. At Darwin's time, some of the facts Johnson concedes were known. Thus, Darwin attempted to explain the progression of life forms, the facts, over time. His initial untested theory, a hypothesis, has had confirmation, but no new species have yet been produced in the lab. This does not fail it as a theory because the time frame for attempting to create new species in the lab has been relatively short. Consequently, evolution is a fact while Darwin's explanation is a viable theory, which accounts for the evolutionary facts. The detailed sub-theories are still in progress.

Additionally, Johnson makes the specific allegation that Darwinism is a "natural" philosophy. But, let us consider the opposite of natural. For instance, in the opposite to natural, if a person drops his or her car keys, it is no longer natural forces causing the keys to travel downward, it is God or the Man in the Moon or Santa Claus or whatever any individual or group wants to *assume*. In Johnson's tangled world, cause and effect is no longer natural. There is no algebra or calculus to measure cause and effect. Four plus one only equals five when God wants it to. Let it be clear, the ramification of Johnson's challenge to "natural" is a delusion of fiction.

What other arguments does Johnson present? In regards to fossil evidence, he cites the lack of transitional links in the skeletal record. This may be true, but there is a possible natural explanation accounting for this fact. The first and most obvious is that biological entities degrade over time when exposed to normal conditions. When millions and billions of years are involved, it is easy to understand very few fossil remains will survive to the present for observation.

But if so, why have so few transitional fossils been found relative to the quantity of non-transitional fossils found? This, too, may be true, but Johnson concedes on page fifty-two of his book, "punctuated equilibrium as developed by Gould and Eldridge" where "the important evolutionary change occurs only among the isolates, who rejoin the stable ancestral population suddenly after forming a new species" (Johnson, p. 52). To clarify, in a large interbreeding population, any successful mutation is mathematically diluted away by the overwhelming numbers of a large herd. In order for a successful mutation to take effect on a population, a small herd size is required. This occurs when small numbers are isolated from the general population. At some future time, the small, mutated population then rejoins the larger population, allowing the mutation to spread mathematically. If this sub-theory is correct, very few transitional links would be expected to appear in the fossil record. This explanation would account for the relatively small number of transitional intermediates actually found.

Nevertheless, the real problem for Johnson's argument, is he concedes, on page forty-nine, "a good sequence of the horse line" (p. 49). That is, in this case transitional link fossils do exist. He's admits so, but never explains this fact away. And clearly, the God theory fails to explain why any intermediates are necessary when an Almighty Creator is involved.

As you may recall, Johnson additionally acknowledges the existence of "Archaeopteryx – an ancient bird with some striking reptilian features." Later, he makes the argument this fossil is probably a dead end in regards to any currently existing species. If so then how intelligent is an assumed God who creates a failed species? If the answer is, God works in mysterious ways, let it be clear such a statement is a concession to – the evidence doesn't support the God theory.

Johnson also makes a seemingly intriguing argument about the advent of organ systems. He states the argument as follows: "Many organs require an intricate combination of complex parts to perform their functions. The eye and the wing are the most common illustrations" (p. 34). To clarify his perspective, what good is an eye without a special eye processing brain part and necessary connecting wires? Consequently, the system required evolving together when the younger species was small and primitive in complexity. Consequently, this argument does not fail Darwin's natural-selection theory. If the example species in question was large and complex without an eye system, and then the offspring species suddenly possessed sight equipment his argument would have some merit. But that is not the case.

So what is said about the human species in Johnson's book? At the bottom of page eighty-one and top of page eighty-two he acknowledges "Australopithecus afarensis, A. Africanus, Homo Habilis, Erectus, and H. Sapiens." Please know "H. Sapiens" are in fact humans. Rather than explain away in any detail each of the first four examples of potential predecessors to humans, he instead chooses to describe anthropology – the study of human origin as "a field that throughout its history has been more heavily influenced by subjective factors than almost any other branch of respectable science." (Johnson, p. 82) Therefore, no matter what evidence is presented in regards to a possible pre-human species, Johnson has already explained them away with the "subjective" nature of anthropology. But if Johnson's argument were to have any validity, he should have argued each of the presented examples away. The fact he did not speaks volumes as to the actual weakness of his position.

In regards to Neanderthal and Cro-Magnon man, Johnson states in small reference print at the bottom of page eighty-two, "Readers who learned about this subject in school may be surprised to find out that Neanderthal man is frequently considered a subgroup within our own species and Cro-Magnon man is simply modern man." He is correct. In fact, in the May 7, 2010 *Journal of Science*, geneticists have determined 4 percent of currently existing Caucasian make-up is from Neanderthal. Consequently, consider the God theory. Does a failed human "subgroup," such as Neanderthal, conceded by Johnson support the Intelligent Creator God theory? Clearly, it does not. Rather, this fact supports Darwin's natural-selection theory.

Finally, Johnson argues with an entire chapter in his book that "Darwinism" is a religion. But Darwin's explanation accounts for the facts of evolution, which Johnson concedes, so it is a theory about a reality issue, despite Johnson's critique, not a religion.

Whereas religion, the Almighty, Intelligent, Creator, God theory fails to account for said facts. And that is exactly why Johnson does not produce a chapter in his book presenting the God theory. Likewise, the intelligent alien theory also fails because it does not explain the gradual life build-up from micro bacterium requiring billions of years. Consequently, only Darwin's "natural selection" theory remains viable as an explanation accounting for the fact of evolution which includes the human species.

Let me make this perfectly clear for those who are still confused. The problem for the religious is not Darwin's explanation of how biological evolution occurred. The problem is the fossil record evidence accumulated over eons, starting with bacterium so small as to require an electron microscope for observation. The observed evidence, including predecessors to humans, conceded by Johnson, is acknowledgment of the evolution. Thus, the observed evolution is fact, not theory! The theory is how the factual evolution occurred. And clearly, the Almighty, Intelligent, Creator, God theory fails abysmally to explain the conceded facts of that evolution.

In conclusion, it is ironic while failing to create doubt about Darwin's explanation, that Johnson inadvertently substantiates this author's claim that the demarcation of philosophy versus science makes no sense. Thus, the academic's have caused the confusion with their philosophy versus science, instead of stating correctly that the issues are reality versus preference.

Chapter Three

Our Real Motivations Resulting from the Fact of Evolution

As a result of the *fact* of human evolution, certain motivations are inevitable in our behavior. Please consider if our ancestors had perished before reaching a reproductive age, no off spring would have been produced, and our very existence would have been in question. Therefore, the motivation for self-survival must somehow be instilled. Secondly, if our ancestors had failed to mate, no reproduction would have occurred, again causing the cancellation of our existence. Consequently, self-survival and species-reproduction must somehow be instilled in our motivation.

Though the human species has evolved considerably, the self-survival and species-reproduction motivations are nevertheless inherent to our genetic makeup. Technically, these are not behavioral instincts. For instance, a bird does possess behavior instincts. Without learning through experience, a bird will weave a nest in a very specific manner. This "how-to" knowledge is genetically inherited. However, the human species has inherited no how-to instincts. This is a well established and documented fact. Nevertheless, we have inherited the two primary motivations shared by other living creatures. But if we eliminate the instinctive or how to components, all that remains is the motivation for self-survival and sexual stimulation. In the human species, any offspring nurturing or group cooperation is always a result of a learned event.

Now let us consider a timeline of evolution. Chronologically, before our current brain makeup, a reptilian brain preceded our mammalian-cortex brain. That is, the cerebral neo-cortex brain found in all mammals, including man, actually evolved over and around the older reptilian brain. It is essential to understand the old brain did not wither away entirely as the new brain evolved. Our older reptilian brain is comprised of the brain stem, the limbic system, and basal ganglia. But why did this second brain, the neo-cortex brain evolve? Clearly, the new brain evolved because it gave greater self-survival potential to the organism (us). In other words, the neo-cortex brain can best be described as a tool, which benefits the self-survival potential of the organism. As a hand or a foot is a physical tool, the new brain is an intellectual tool.

But, how does the new brain, the neo-cortex brain, differ from our older, primitive reptilian brain? The answer is our neo-cortex brain is an intellectual brain while the old brain is a non-intellectual brain. Let me explain. The old brain operates in a pre-programmed manner. That is, this brain cannot learn. If it were programmed to blink it would do nothing but blink. Our neo-cortex brain on the other hand is an intellectual brain in that it is capable of comparing and memorizing information, thus learning.

Of primary significance is the following: Our consciousness resides in the neo-cortex brain. That is, our sense of self, where we reside, is located in the new brain. Our old brain is completely subconscious. It operates the involuntary self-survival functions such as heart beat. For instance, in the case of severe trauma to the neo-cortex brain, the body often lives on in the vegetable state when the personality is quite gone. Thus, the old brain, programmed by evolution, is responsible for maintaining the vital body functions that support self-survival.

But how could the new intellectual tool, where we reside, be controlled for the evolution created purposes of self-survival and species-reproduction? Consider the obvious. The intellectual tool is controlled by the application of pain and pleasure. These feelings are a result of chemical releases in the new brain. These brain chemicals are called neurotransmitters. To date about fifty neurotransmitters have been identified. The neo-cortex brain is comprised of about one hundred billion neurons that transfer intellectual information, as well as pain and pleasure sensations by passage of these chemicals.

When the chemical caused sensations of pain and pleasure are linked to our intellectual thoughts, we classify them as emotions. Pain and pleasure sensations are the motivation to operate our intellectual capabilities for the evolution created purposes of self-survival and species-reproduction. Consequently, the neo-cortex brain, where we reside, is intellectual yet emotionally driven.

One of the renowned psychologists of the last century was a Dr. Maslow. Maslow suggested there is a hierarchy or priority of needs. The most fundamental priority is that associated to what our blood carries such as oxygen and nutrients. For instance, if for some reason our blood should become oxygen deprived, certain neurotransmitters are released in our neo-cortex brain causing the strong sensation of pain. The pain becomes the motivation for operating the intellectual capacities in the direction that will resolve the pain and thus benefit self-survival. Therefore, the most fundamental priority is established as an avoidance of extreme pain.

Second on the hierarchy of needs, according to Maslow, is the need for shelter. Without protection from ambient cold and heat, we again experience pain sensations. An ensuing chemical neurotransmitter is released causing the motivation for the intellect to resolve the temperature problem, which as before, ultimately benefits self-survival.

As we progress upward on the hierarchy, different chemicals are released, creating the motivation to use our intellectual capacity for the purposes of self-survival. Once the avoidance of pain has been accomplished, as at the bottom of Maslow's hierarchy, the pursuit of pleasure becomes the objective. Again, certain neurotransmitters must be released in order to cause the various pleasure sensations. For instance, one phase of sexual gratification occurs when several chemicals, dopamine, noradrenaline, and phanathamean, are released simultaneously. The resultant pleasurable feeling is the motivation that prompted the sexual activity that, in turn, statistically produces offspring, thus benefiting species survival.

At the top of Maslow's motivational ladder is what he terms the drive to achieve esteem. I prefer the Freudian term, "ego." And as with all human motivation, the ego rush is ultimately chemically induced. The pursuit of these chemical releases creates the incentive to over achieve or surpass basic requirements. But how would this benefit self-survival? Clearly, the surplus hoarded by an animal ancestor served as insurance for that creature when times were lean. Thus, over achievement induced by chemical releases is statistically advantageous to self-survival. In fact, these ego chemicals are the primary source of motivation for the person whose basic survival has been achieved.

In regards to motivation, many different neurotransmitters are released in various combinations in order to motivate us in the statistical direction of self-survival and species-survival. The resulting intellectual feelings we experience are of vast variety. But, all are chemically induced. Nevertheless, at the fundamental level, all human motivation emanates about avoidance of self-pain and pursuit of self-pleasure. However, due to our intellectual capabilities, we often deliberately endure pain in order to obtain a goal perceived to bring pleasure. For example, a dog utilizing his intellectual tool will knowingly endure physical pain in order to obtain food. Human beings are clearly no exception. People often work at jobs they consider miserable in order to obtain a paycheck associated with pleasurable ramifications. Some religious people deliberately abstain from many physical pleasures (which causes internal chemical releases) in order to obtain their perceived reward at the end of life. Clearly, the ultimate goal is pleasure, and everything else is simply a means to this end.

What about the person who commits suicide? How could this action be a by-product of evolution created self-survival? Please remember the process of evolution shows no signs of a carefully engineered and efficient plan. Suicides are proof to the inefficiency. However, the pain and pleasure chemicals do motivate us statistically in the direction of self-survival and species-survival. Please note: the current world human population is about seven billion and climbing. When people do commit suicide, the motivation is either avoidance of pain or pursuit of pleasure in a perceived after life.

In summation, our ultimate motivation is avoidance of self-pain and the pursuit of self-pleasure. Had the processes of evolution been different, for instance if these processes had caused the release of the pleasure chemicals when we banged our heads against a wall, then that would have become our goal? That is, we would all be deliberately banging our heads against walls in order to cause the release of the pleasure chemicals. As proof, please consider the damage frequent drug users do to themselves as they pursue the pleasure of narcotic chemicals. Nevertheless, avoidance of self-pain and the pursuit of self-pleasure have statistically benefited self-survival and species-survival, causing the ballooning of the human population. This additionally explains why preference issues exist.

Finally, I must state the intellectual tool (the neo-cortex brain where we consciously reside) has no built-in allegiance to the truth or that which is consistent with the facts. Furthermore, the intellectual tool has no built-in concern for others. Rather, an allegiance to the truth and a concern for others is always a result of a learned event.

Chapter Four

The Failed God Theory and Why It Is Promoted

From chapter one of this book, I have established that there are no philosophic issues; there are no scientific issues. Again, as fact, those are methods, not issues. To the contrary, as previously stated, the actual issues are those about reality and those about preference. And any issue that infers something exists is a statement about reality, not a stated preference. Consequently, like it or not, the promoters of an All-mighty, Intelligent, Creator or "God" are entranced in a reality issue. And how is reality understood? Theories must explain facts. And because those promoting religion have failed to prove the existence of an Almighty, this reality issue automatically degrades from proven fact to conjectural theory.

So, why is it being promoted as a faith issue? How else can something that does not actually exist be promoted? For instance, if someone attempted to sell an automobile that does not actually exist the customer must first be encouraged to accept on *faith* that it does. Bear in mind, if the customer asks for any direct proof, none exists. Consequently, the promoter must first convince the customer no proof is required. This explains why religion is promoted, by the promoters of religion, as a faith issue. But let us distinguish fact-based confidence from blind faith. For instance, leaping out of an airplane in flight with a parachute would exemplify fact-based confidence. That is, parachutes are proven as reliably safe. On the other hand, leaping out of an airplane in flight with only a Bible would exemplify true faith. Now, ask yourself if those promoting faith would be willing to make the Bible-only jump? Consequently, a significant difference exists between fact-based confidence versus blind faith. Let me make this clear. The sane operate upon fact-based confidence whereas blind faith is a deceptive promotion technique used to control the confused.

In an attempt to overcome this confusion, let us return to the consideration that the God theory, as with any theory about reality, must be supported by facts. The fact that we, the human species, and the universe do actually exist has been used by the religious in order to rationally justify religion from the beginning of religion. But chapter two of this book dispels an All-mighty, Intelligent, Creator, or God was required for the production of our species. But, what about the creation of the known universe? Is an All-mighty necessary for such a creation?

If the religious argue, what would have started the beginning of the universe if it were not for a Creator? Consider the obvious: there is no absolute beginning. After all, if a Creator started the universe, then who started the Creator? The only intelligent answer is, there is no absolute beginning. Rather, the system that created the known universe must be infinite, whether a Creator is involved or not.

But if no Creator was involved, then how could the universe come about? Please consider Albert Einstein's famous mass energy equivalence equation: $E = Mc$ squared. The equation implies mass is derived from energy. As fact, this transformation occurred during the formation of the universe as in the big bang. In the Timothy Ferris book, *The Whole Shebang*, on page thirty-two he states, "What we call matter is frozen energy. It froze because the universe, owing to its expansion, cooled." Let it be understood the reverse, mass to energy, as in sun produced light also occurs. In this case, the hydrogen "burning" to helium equation is, 4H + 4e > 4He + 2e + 2v + 26.7MeV. Please note as fact the total mass of particles on the right side of the equation is less than that on the left side. Thus, mass is transformed back into pure energy by the gravitational pressures created in a star. Furthermore, the space between galaxies and all constituent mass has been measured to be expanding at an accelerated pace. A "dark" or transparent energy seems to be reducing frozen mass towards an inevitable return to pure energy. A reasoned conclusion or theory: It is likely mass and energy cycle back and forth infinitely.

Recently, the physicist Stephen Hawking suggested because time *might* stop in a black hole, the initial state of the universe similarly would have no time. Combining this with quantum small scale physics, he suggests, no preceding event is necessary for the creation of a universe. Other researchers are currently searching for plausible explanations or theories, which account for the beginning of this universe while acknowledging the probability of other universes. In conclusion, though the universe seems to be complex in detail, Einstein's famous equation proves otherwise. In any case, no Creator is required in the birth of a universe. It is just the event of energy cycling to mass.

Another argument made by the religious is the fact the natural world is complex, and yet, everything seems to fit precisely in order to support life. Therefore, random luck must be ruled out in favor of an intelligent Creator. But no one claims random luck in regards to the process of evolution. For instance, if a gene code is randomly altered or mutated, the subsequent, altered creature will survive only if that creature fits the environment. Thus, the actual survival would not be luck. In fact, the vast majority of mutations do not survive. Consequently, the process of evolution is only partially random.

Also, a deliberately misleading argument (originated by some in the educated religious community) suggests the laws of entropy are violated with the organization of earth's biological complexity. This is false because the laws of entropy allow for complexity when energy is being added to a system as occurs from sunlight radiation to earth.

In general, those searching desperately for evidence to support the failed God theory look for any gaps in research as evidence of a Creator. To the philosophy department's credit, they recognize this desperation with the terminology, "God in the gaps."

But how do we know an All-mighty is not somehow involved? In addition to previously stated facts in chapter two that contradict the theory of an "All-mighty Creator," a question must be answered. Where is he? As explained, this is not a test of faith. Therefore, a God has no reason to hide.

What about the authority of the Bible? "It can't be challenged" is often stated in order to promote certain religions. But if you open a copy to the first page the human printer is named. So, the fact is, a Bible is printed by humans. Furthermore, as fact, it was translated by humans. And of course, it was originally written by humans. So, why is it that the Bible cannot be challenged? After all, if something is true, thus consistent with fact based reality, why fear an intellectual challenge? To the contrary, it is *only* when something false is presented as true that fear of a challenge is certain. Thus, the promoters of religion attempt to prevent any challenge to their manmade Bible.

Finally, when someone presents something as true or consistent with reality, the burden falls upon the presenter to produce facts which support the argument, the theory. Because the All-mighty or God theory has failed to explain known facts, those promoting it defaulted the issue to plan B, and sell it as a faith issue. But, they themselves don't believe it. So, what is their real motivation? How do they benefit from promoting religion, a failed theory about reality?

Please consider the stature gained by anyone controlling large numbers of society. This would include the leaders of religion. As an example, the religious leader Billy Graham was a frequent guest in the White House. As fact, he often golfed in exclusive country clubs with those elected officials. Advanced scouting was usually employed in order to secure his "proper" hotel stays. In many categories, his life style could be equated to that of a musical rock star. All this happened because in the modern Western world controlling large numbers in society means controlling large blocks of voters with a subsequent election influence. And in concept, the ancient world was no different. Those promoting religion could benefit those in power and those in power could in turn benefit those promoting religion. Thus, politics and religion sleep together.

But what about the vast majority of those involved in religion? What is their motivation? To their credit many of those involved in religion, as fact, give help to many of the poor and indigent. Clearly, these are non-selfish deeds resulting from admirable intentions. As for the motivation of the follower, religion's false promises of eternal life aligns with our evolution caused motivation for self-survival. It's an easy sell to those unconcerned with facts. Additionally, religion's heaven or hell aligns with nature's pleasure and pain neurotransmitters causing our fundamental motivations (refer to chapter three of this book). Therefore, man's invented, thus fictional, heaven and hell is the ultimate psychological carrot and stick approach to controlling the human herd. This is enforced by the authority of a non-existent All-mighty, Creator, God.

However, for those concerned with the truth, the facts do not support the existence of a Creator. Religion is merely man-made superstition invented for the purposes of controlling the human herd.

Chapter Five

The Preference Issue of Morality

By definition, morality is principles of right and wrong in conduct. But who makes the judgments of what is right and wrong? To the religious right, judgment of conduct is an absolute made by an All-mighty, God. But upon analysis of directly related facts, no such Creator actually exists. To the liberal left, judgment of conduct is relative with no absolute. But as stated in chapter one of this book, morality is relative to someone's preference. Therefore, morality is a preference issue. However, this author would argue no one prefers to be wronged and whenever anyone thinks he or she have been wronged, a lack of fairness is the perceived cause. Consequently, all rational judgment of conduct (morality) is dependent upon fair treatment to all. But let us review the definition of the word "fair," so we eliminate any potential misunderstandings.

By definition, fair is all sides treated equally. But let us add a necessary caveat. It is choice dependent. That is, for fairness to occur all sides must have an equal choice. Let me elaborate. In the United States, before the Civil War, black Americans were enslaved by no choice of their own doing. That is, they were denied equal (fair) treatment by law. Compare this to someone who is jailed because they choose to commit a crime, which impacts others. In both situations, freedom is denied, but in the latter case, it is fair because it is a consequence of a choice. That is, anyone who chooses to commit the crime would be equally punished. In the former case, no choice was available to the black Americans. They could not choose to be white slave owners. The enslavement was brazenly unfair. Thus, fairness is choice dependent.

Additionally, we must distinguish man-made from nature-made. For instance, have you ever heard someone say "life is unfair?" The problem with this ambiguous expression is that it fails to distinguish between what man does versus what nature does. Clearly, in regards to conduct, it is only what man does that is in question. Keep in mind, the hypocrite who says "life is unfair" will not except being cheated by others.

So, what is the opposite to fair? Self-advantage for some, which causes a disadvantage to others. It is ironic that many in society would consider themselves moral while promoting self-advantage over others. Likewise, it is ironic in the United States that the legal system is called the "justice system." By definition, justice is the quality of being impartial or fair. That is, the words justice and fairness are synonymous. But clearly in the past as indicated by the enslavement period of black Americans, the system was extremely contrary to justice. That is, laws can be passed with legal enforcement by the justice system, and yet, they are in obvious contradiction to justice by definition. Today's gay population has experienced this truth as a result of religious superstition combined with religious voting potential.

But if morality isn't based upon justice (fairness), then what should it be based upon? To the religious, morality is found in a book printed, translated, and written, as fact by man, but purported to be inspired by an intelligent Creator. But the facts don't support the God theory. Furthermore, it is remarkable that those of one religion often believe all other religions are false. Thus, this self-serving perception allows for the dominance of the "true" believers over all others, the infidels. Nevertheless, for the rational, fact-based person, morality established in a religious book is nothing more than man-made superstition.

In conclusion, morality is a preference issue as opposed to a reality issue. To the liberal left, morality is relative. But as fact, morality is relative to someone's preference. However, the question becomes, *Is someone's preference a desire to be fair, or is someone's preference a justification for gaining self-advantage over others?*

Chapter Six

What is Conservatism? What is Liberalism?

What do kings, czars, emperors, and sheiks, all have in common? By right of birth, theirs is an inherited, superior position in society. But how can such a privileged unfair entitlement be maintained? Please, consider the conservative philosophy, or more correctly, from chapter one of this book, the conservative preference.

By dictionary definition, a conservative is someone who tends to preserve established traditions or institutions. Historians attribute modern conservatism with Edward Burke's (1727-1797 AD) preference to resist catastrophic social upheavals as he cited was caused by the French Revolution. But how does a modern conservative explain conservatism?

An essay entitled, "The Conservative Agenda: Its Basis and Its Basics", written by Norman L. Geisler and posted on line cites the American Declaration of Independence from which he deduces, "The most fundamental principles of conservatism are the first three: Creator, Creation, and God-given moral absolutes. These are the foundation of our country, our constitution, our courts, and our conservative agenda." However, it is extremely ironic that he cites the Declaration of Independence as a document for the foundation of conservatism in America because that very document is a statement giving the reasons to break away from an established institution, the British government. From the

Declaration of Independence: "It is the right of the people to alter or to abolish it, and to institute new government." Clearly, this document is in dire contradiction to the definition of conservatism. Thus, Norman L. Geisler is very confused. But please note, of the entire document he somehow interprets "Creator, Creation, and God-given moral absolutes" as the fundamental principles of conservatism despite that "governments are instituted among men, deriving their just powers from consent of the governed," as in the Declaration of Independence.

But how does a popular modern conservative explain conservatism? From his book, *The Way Things Ought to Be*, conservative radio talk show host, Rush Limbaugh, states, "I believe in the individual, in less government, so as to allow that individual maximum freedom to create and achieve" (p. 2). Though, this is a statement of preference, it is nevertheless ironic that conservatives tend to prefer the promotion of law and order, the opposite to individual freedom. For instance, it is the social conservatives who tend to fight the legalization of marijuana, abortion, or gay rights. This authoritarian conservatism is diametrically opposite to the stated libertarian conservatism that is sold.

But why would conservatives prefer less government? Rich conservatives such as Rush Limbaugh, Donald Trump, the Koch brothers, etc., are born into wealth. They have no need for government social services. Thus, keeping his or her own taxes low is the *primary* conservative agenda, as established by the *rich* conservatives. But in order to deceive the less fortunate, low taxes for government revenue is deliberately misrepresented, in Limbaugh's own words, as "less government so as to allow that individual maximum freedom."

Limbaugh goes on to say, "That society owes its citizens equality of opportunity, but cannot guarantee them equality of outcome" (p. 2). However, this, too, is in conflict with actual conservative agenda. For instance, as pushed by conservative Republicans for 2011, the offspring of the wealthy shall be entitled to receive five million dollars federal tax-free inheritance from each parent for doing nothing, while those not born to such entitled wealth will be required to pay federal income taxes at the normal rate, while producing goods and services benefiting society. Clearly, the actual conservative agenda is opposite to Limbaugh's published statement.

Limbaugh continues, saying, "That there is one God, and that this country was established with that foundational belief; that our morality emanates from our Divine Creator, whose laws are not subject to amendment, modification, or rescission by man" (p. 2). But this conservative belief, that a God exists, conflicts with fact-based reality, despite the confusion created by the philosophy department, "It's a philosophic issue." So, why would conservatives continue to promote an irrational belief inconsistent to hard evidence? Please consider any unfair laws created by conservatives: "Are not subject to amendment, modification, or rescission by man," in Limbaugh's own words. Therefore, it is not self-serving men who create self-serving laws; it is God who creates just laws. And we know a mere human cannot challenge the authority of God.

Furthermore, on page two of Limbaugh's book, he states that "human life is sacred and that God placed man in a position of having dominion over nature." Again the God-existence issue conflicts with fact-based reality. Clearly, to make such an assumption, despite the facts conveniently authorizes any violations to the animal world, our genetic relatives. Consequently, according to conservatives, animals are placed here deliberately by God for man's consumption.

Additionally, most American conservatives are quick to cite the authority of the US Constitution. They often say, "It is in the constitution." From which we can take, therefore, it can't be challenged. But again, the irony of the conservative preference is their ubiquitous focus on the Second Amendment, the right to bear arms. What makes this contradictory to the authority of the Constitution is that an amendment is proof of the original Constitution's imperfection-something that needed to be reworked. And ironically, for conservatives to cling to an amendment, as they do, this is a complete contradiction to the very definition of conservatism, which says, don't change or amend. Clearly, conservatism defined by conservatives is confusing, false, self-conflicting, but mostly, self-serving.

But are conservatives fiscally conservative? And is fiscally conservative the same as fiscally responsible? Recently, the Republican legislative conservatives negotiated the extension of George W. Bush's tax cuts for those Americans with incomes exceeding two hundred and fifty thousand dollars a year, despite their complaint of government indebtedness. If being fiscally responsible means balancing a budget the tax cut extension was contradictory. What fiscally conservative apparently means is reduced taxation for the wealthy, not to be confused with fiscally responsible.

Additionally, as previously stated, conservatives in America speak of a less (smaller) government preference. But, this, too, is deceptive. For instance, under conservative President Ronald Reagan, the military budget was increased substantially. As a consequence, US debt tripled under the eight-year Reagan administration deficit-spending policy. Consequently, it is extremely ironic that modern conservatives hold up Reagan as the model of modern conservatism.

Let's be clear, what conservatives actually mean by "less government" is a less social services government. Consequently, modern conservatives often refer to social programs like social security as unearned "entitlements." But when someone pays into an insurance program and collects (without fraud involved), it is earned. For instance, if a person buys auto insurance and the insured vehicle is stolen, the money received for the loss has been earned because of premium payments made. Likewise, social security premiums are deducted from earned payrolls. Contrast this to the unearned birth entitlements received by the offspring of the wealthy. The offspring don't earn the money. It is a birthright "entitlement" enforced by manmade law.

But do conservatives own legitimate claims to morality? Are they the moral majority? If fairness and truth (that which is consistent with fact-based reality) have nothing to do with morality, then morality can be anything, and anyone, including conservatives, can claim ownership. But if fairness and truth are of preference, then the conservative preference is without morality. In fact, both fairness and truth quickly evaporate as conservatives strive to maintain the status quo.

Let us review what conservatives emphasize. They emphasize the law, God, and guns. But, what is in common? It is about their authority and don't challenge it. But if the smoke and mirrors are removed, the preference to maintain an unearned birth status remains. Let us be clear, the conservative preference has been around long before the American, Edward Burke. The Russian czars, the Japanese emperors, the sheiks of the Middle East, and the pharaohs of Egypt are all examples of conservative preference. And the myth of a God authority usually sanctions the preference.

So, what is liberalism? In the context of conservatism, a liberal is someone favoring reform or change. Unfortunately, the change isn't always for fairness. In fact, in regards to fairness, there have been some legitimate counter arguments made by the conservatives. For instance, liberal affirmative-action policy has favored some unqualified minority members over qualified white majority members. Clearly, in certain circumstances, the affirmative action policy has been unjust. Additionally, in the past, welfare was a reward for some to produce nothing. Ironically, it was during the moderate, liberal Clinton administration that the welfare program was reworked in order to reduce life-long welfare dependency. Nevertheless, liberal policy has rewarded the undocumented with tax payer emergency health services where the costs have been unfairly absorbed by legal citizens. And liberal teachers' unions legalized tenure, preventing inept teachers from employment termination.

Taken to the extreme, far left liberal preference has been associated with Communism while the hard right engenders capitalism. By definition, Communism is the economic system in which the ownership of all means of production are held by the community with all members sharing in the work and the product produced. But without individual incentive or motivation to create, product is limited, and what gets created is distanced from actual consumer needs. Thus, as evidenced by the old Soviet Union, communistic failure was all but inevitable.

By definition, capitalism is the economic system in which all or most of the means of production are privatively owned and operated for profit, originally under fully competitive conditions. To clarify, the words "fully competitive conditions" implies all sides being treated equally and are primary to the efficiency of any economic system rather than the word "capital." Otherwise, individual incentive is reduced when the component of fairness is limited. As an analogy, in a foot race, someone starting too far back will lose incentive because of a limited potential to succeed, while someone starting too far ahead will lose incentive because winning without effort is guaranteed. Both situations produce less than maximum effort thus creating inherent inefficiency. Historically liberalism has rewarded the former with welfare while conservatism has rewarded the latter with near tax-free inheritance.

In summation, conservative and liberal preferences have failed to focus on fairness. Thus, less than maximum economic efficiencies have resulted. The consequential impact to others shall be presented in the next chapter.

Chapter Seven

Who Gets Rich, Why, and the Impact to Others

Economics is a reality issue explained by theories accounting for facts. Like other reality issues, theories are often expressed as equations linking the factors together. Additionally, people called economists link the factors of price, supply, and demand into two-dimensional graphs called supply-demand diagrams. With that in mind, let us determine who gets rich, why they get rich, and the impact this has on others.

As an analogy, please consider the following: If there is a twenty-six mile marathon foot race, where person A starts at zero and person B starts at plus twenty miles, who will win the race? Clearly, everything else being equal person B should overwhelmingly win. This is because cause and effect is measured with an algebraic equation where more than one factor is involved. If all the other factors are equal, than the outcome is certain. In fact, if person B fails to win by at least a twenty mile margin, that runner is less competent than the losing runner.

Now, please consider the economic world. Donald Trump is known to be a successful and wealthy land developer. According to Trump in his own book, *The Art of the Deal*, his first building purchase, using his dad's money and signature was for millions of dollars. As it was, the senior Trump was a financially successful multi-millionaire (billionaire in 2014 terms) land developer. If we consider the opportunity factors involved for the young Trump, they would include, a very successful mentor who explained pertinent business aspects not taught in school, business contacts at the highest level, the actual money loaned by the mentor, as well as the mentor's signature on the loans. Combined, the opportunity factor for Trump, through nepotism, was vastly greater than the opportunity factor for the average person. In fact, if Donald Trump's financial success was not far greater than that of the average person, his competency would be in question. Please understand that the average person with a college degree does not receive millions to invest. Now consider conservative Rush Limbaugh's written words from his own book, *The Way Things ought to be*, "That society owes its citizens equality of opportunity." Clearly, Limbaugh is grossly divorced from factual reality if he thinks there is equality of opportunity. Nevertheless, is it any wonder the rich get richer?

Next, please consider someone at the other end of the spectrum. Let us consider Ray Crock, founder of the McDonalds food chain restaurants. Mr Crock apparently started with no special opportunity factor. In fact, in his quest for success, Mr Crock filed bankruptcy four times, but, his apparent motivation was ultimately achieved. These two extreme cases (Trump and Crock) exemplify opportunity and motivation. The equation can be expressed as success equals opportunity multiplied by motivation, where opportunity is that which is outside of an individual's control (luck) and motivation is that which is within an individual's control.

But what is the impact to others? As per basic economics, a demand curve for a given product goes down as prices go up. Conversely, a supply curve for a given product goes up as prices go up. Accordingly, the demand is in equilibrium with supply at the intersection of a supply-demand diagram. Again, as per basic economics, if consumer income is increased without altering supply, a shift to the right occurs in a supply-demand diagram. The corresponding price consequently goes up. Or, think of it this way, when the ratio of purchasing power or demand, to a fixed supply goes up, an increase in price occurs. Consequently, in a free market, price is proportional to demand and inversely proportional to supply. This relationship can be stated as, price is proportional to the demand-to-supply ratio. This is the governing relationship for an economic system.

For instance, if there is a reduction in oil reserves (supply) for any reason, with an established demand the price per barrel goes up because the ratio of purchasing power (demand) to supply goes up. If a government prints additional money or reduces interest rates, more money chases supply, and the ratio predicts prices will go up. If supply is artificially controlled by a monopoly the price to the consumers also goes up. Additionally, this inflation can be an already added cost, one unknown to the consumer because of deliberately reduced supply or increased unknown artificial demand such as that caused by speculation buyers.

Now, let us consider what happens to price when someone gets rich. If an individual gets rich by causing increased supply, with no monopoly involved (fair competition), the ratio of demand to supply has not been adversely affected. In this case no inflation occurred. More supply was produced. No sane person should have a complaint. In the case of Ray Crook, this seemed to be so. Trump, on the other hand, did not have to compete for his position with his dad. The lack of real competition inherently caused some increase to the purchasing power to supply ratio. Additionally, it is extremely ironic that he is known for his statement, "You are fired," when he was never in that position with Dad. Bill Gates made his fortune with the ownership of the PC operating software. Many lawsuits were filed against Gates for monopolistic activity. If the allocations were true, then the consumer paid more because price was increased due to a decrease in supply resulting from a monopoly.

For those who get rich strictly through inheritance, the maximum inflation (cost to others) occurs. This is because no goods and services have been produced by the recipient in order to receive money. Thus, the purchasing power to supply ratio has increased, causing others to pay more. It is ironic the conservatives want this money to be taxed at a lesser rate than the rate for producing goods and services. Clearly, the conservatives want everybody else to compete tooth and nail, producing maximum supply while they receive at the minimum tax rate for producing nothing.

The conservatives do have a canned response. Their counter argument goes as follows, person one already paid the taxes on the earned money; therefore, the gift of that money to another should be without additional taxation. But if this is their preference, then why should someone else not be able to give his money to a contractor, with the contractor, in turn, giving his services back. Thus, all money could be "given" without taxes ever paid. Instead, the conservatives want taxes paid for producing goods and services while preferring no taxes on inheritance (nothing produced). Their preference is the diametric opposite to economic efficiency, but it benefits conservatives born into wealth. Of course, this will never be presented on the right-leaning Fox News network or the conservative Rush Limbaugh ratio talk show.

At this point, let us consider the difference between a Robinson Crusoe, isolated island versus an economic interconnected system. The governing relationship for an economic system, price is proportional to demand to supply ratio, is not affected outside of the isolated Robinson Crusoe island. Therefore, what happens economically there has no impact outside that system. However, in an economic system, as we live in, the demand to supply ratio holds because the system is interconnected. This means as an individual, you have more buying power if your competitors have less money and less buying power if your competitors have more money. This factual reality is experienced when someone else out-bids you for a house you wanted. But if they have produced at a competitive rate, no legitimate complaint is valid.

Nevertheless, besides inheritance and shoe-ins by Dad, many CEOs receive excessive pay allowed by members of the board because board members are often either chosen by the CEO or are approved by the CEO. These consequential crony relationships lead to excessive executive salaries. And the current conservative-dominated Supreme Court in the United States recently ruled that corporations can legally contribute unlimited funds to political campaigns with their Citizens United decision.

Consequently, the United States is a "pay to play" system, which favors those with money. It functions as follows: The primary motivation for a legislative Congress person is re-election, which benefits his career. The probability of re-election is increased by media, such as television advertising, and paid campaign staffers. Campaign contributions pay for much of that advertising expense, as well as the professional staffers. Those writing the checks will go to individuals of Congress offering policies that favor those writing the checks. Consequently, the United States democracy is not a pure "democracy" with all sides having an equal voice. Though various groups, such as political parties and unions also contribute to campaigns, nevertheless, the offset still favors individuals with money as opposed to individuals without money. Consequently, tax loopholes permeate the tax code and primarily benefit the wealthy.

Now, let us consider the purpose of an economy. The purpose of an economy is to produce the maximum amount of needed goods and services. The purpose is not to increase job quantity as some might think. For instance, if there were an infinite amount of goods and services available, no one would have to work. That said, work is needed in order to produce the supply of goods and services, but *needed* goods and services is the goal, not jobs. For instance, a job that produces unneeded goods and services is highly inefficient. Some government contracts produce unnecessary product. Recently, the Raptor fighter jet engine, unnecessary redundancy, was finally cancelled. That contract produced jobs without producing wanted and needed product. But the conservatives suggest a "real" job is only in the private sector where true competition exists.

However, the factual reality is many private-sector industries exhibit monopolistic tendencies. For instance, the health care, poultry, and pharmaceutical industries have made many wealthy because of the lack of real competition. For instance, the average CEO in health care makes about seven million dollars a year. Is it any wonder, during the period when President Obama was attempting health care modification, the public option was vehemently denied by those benefiting from the status quo? Competition from the public sector was the last thing their "free market" actually wanted. Additionally, forty percent of all campaign contributions to congress (Republicans and Democrats) come from the banking industry. The industry is paying for favorable laws that prevent real competition. But the world's most profitable industry (in terms of largest quarterly earnings) has been that of oil production. Ironically, as a result of lobbying efforts and campaign contributions, the United States tax revenue actually subsidizes this highly monopolized and profitable pursuit. Clearly, profit from oil revenue has made some very wealthy.

During the 2007 economic slow down, the conservatives successfully extended the 2003 Bush tax cuts (despite President Obama's reluctant approval signature) for those with incomes greater than two hundred and fifty thousand dollars per year. Those with incomes below two hundred and fifty thousand dollars were unaffected by the tax cuts. The Republican justification for the tax cut extension was to free up the "job creators'" income in order to allow the hiring of more job seekers. This, despite the fact the 2007 recession occurred with the tax cuts fully in place. The trickle down, economic, self-serving theory continues to fail at producing consumer demand, but it partially accounts for the uneven wealth distribution in America. Incidentally, benefiting from the tax code, thick with deliberate loop holes, is ex-Republican presidential candidate Mitt Romney, son of millionaire George Romney. As publicly disclosed, he paid just under fourteen percent in federal taxes for 2011, due to the lower capital gains legislation, which allows the rich to game the system.

Finally, according to an on line article, "Who Rules America?" by professor G. Williams Domhoff, in 2007, one percent of the U.S. population owns 42.7 percent of the nation's financial wealth. Also, the richest four hundred individuals own more than the bottom fifty percent.

Chapter Eight

Why the Philosophers Are Wrong!

Now that you, the reader knows there are issues about reality and there are issues about preference, how does this compare to the perspective of the so-called "great" philosophers? The first thinker to be associated with Western philosophy was Thales (635 -543 BC) from the city of Miletus in Greece. Though he was essentially a mathematician, Thales is especially known for his so-called philosophy "all is water." But is this a philosophy? Clearly, "all is water" is no more a philosophy than Einstein's famous "energy equals mass multiplied by the speed of light squared." Both are in fact theories about reality as opposed to stated preferences. And both theories are testable to a certain degree. But the similarities go farther. Let it be understood Einstein's transmutable equation implies "all is energy." As presented in chapter four, Ferris states, "What we call matter is frozen energy. It froze because the universe, owing to its expansion, cooled" (p. 32). Consequently, Thales was incorrect. Furthermore, Einstein's "all is energy" has enormous implications to explaining where everything comes from.

Socrates (469-399 BC), like Thales before him, wrote nothing, but we know of his thoughts through the writings of Plato. Accordingly, in Plato's *The Apology* much of Socrates' "philosophic" thinking is presented. Socrates was primarily concerned with ethics and morality, which are preference issues. He believed in doing what is right in the face of universal opposition and the need to pursue knowledge even when opposed. Though many people might agree, these are nevertheless stated preferences. And, what is considered "right" is most often culturally learned with no absolute other than for the caveat of fairness. But, Socrates is additionally known for the Socratic' method of working through an argument in order to expose contradictions. In chapter one of this book, I stated we determine what is true and consistent with reality, by comparing information. Clearly, the Socratic Method relies on comparison. Otherwise, contradictions are missed. Consequently, it is extremely ironic. This book exposes the faulty demarcations of philosophy versus science as issues based upon the method produced by a founding father of philosophy.

Plato (429-347 BC), a profound writer and student of Socrates is generally considered a dominant figure in Western thinking. Much of his work was written as dialogues. In them, Plato suggests that beyond the imperfections, as appears to our senses, is a perfect form. This duality occurs, according to Plato, with issues such as goodness, beauty, geometry, and change, to cite just a few. The problem with this thinking is the failure to separate out issues of preference (goodness, beauty) from issues about reality (geometry, change). Beauty, for example, is in the eyes of the beholder. One person's beauty is often another's ugly. Therefore, no one perfect form can exist for any preference. To the contrary, geometry such as a perfect straight line is a measurable "reality" issue. And, no real line is ever perfect. In this case the duality holds, but so what. This duality answers no significant questions.

Aristotle (384-322 BC) studied under Plato, but often disagreed with him. In his thinking, knowledge was empirically derived. Like modern science, Aristotle thought that through *inductive* reasoning it was possible to determine the principle (a general-case theory) from related evidence. Aristotle went on to suggest the truth is determined by the facts they represent. And this is exactly why theories about reality must be supported by all the available related facts (one of the center points in this book). But induction is no big deal. As argued on page 106 in Will Durant's nineteenth printing of *The Story of Philosophy*, Durant writes, "Induction has been practiced from morning to night by every human being since the world began." (Durant, p. 106)

Additionally, Aristotle classified specific knowledge based upon the relative certainty with which you could know those objects. But today's uncertain information may become certain in the future as additional facts are produced. The other problem with this classification is the failure to separate out human preference. Consequently, the big picture according to Aristotle's classification, is absent without the most fundamental classification of reality and preference. As a result of this confusion, Aristotle claimed a "golden mean" for the pursuit of happiness. But, while this may please some, others will prefer the extreme. Thus, his golden mean is just a stated preference, nothing more.

During the classical era, politics were also vigorously debated. In the Will Durant book, a school of thought similar to that of Friedrich Nietzsche (a nineteenth century philosopher), "Claimed that by nature, all men are unequal, and that of all forms of government, the wisest and most natural is aristocracy" (p. 7). But all men are unequal in what way? Bear in mind everyone has unique abilities. Thus, no absolute superior or inferior actually exists. And anyone who makes the superior or inequality assessment is forced to use his or her own preference for that judgment. Therefore, the equality issue is a preference issue. And the way individuals are treated is a preference issue, as well. From chapter three of this book, I write the brain is an intellectual tool, which evolved in order to resolve one's own pain and pleasure. Thus, for those born into the aristocracy, the intellectual tool must justify (in order to perpetuate) his or her unfair advantages. Consequently, the school of Nietzsche-like thought produced during the classical period is a twisted justification serving the aristocracy, not unlike the current thought produced by modern conservatism.

In the Will Durant book, Durant goes on to say on page seven, "No doubt this attack on democracy reflected the rise of a wealthy minority at Athens, which called itself the Oligarchical Party, and denounced democracy as an incompetent sham." In modern United States, the politicians who will create unfair laws are bought with campaign contributions giving those with money the unfair advantage. The nonsense changes form, but not function. Consequently, no democracy is true to equal voices.

During the Renaissance Period Rene' Descartes (1596-1650 AD) of France, like the Greeks before him, was concerned with knowledge certainty. Descartes is known for his second meditation, "I think, therefore I am," as a base of knowledge certainty. But, how was he certain he was not dreaming the idea, "I think, therefore I am"? Thus, this author would argue no knowledge is completely certain. Rather, think in terms of relative certainty. Consequently, we classify facts as our most reliable information. Therefore, in any understanding of reality, the facts must be addressed. And that is what theories do. Theories account for the most reliable information, the facts. But theories themselves are usually considered less certain than facts. This presents a problem in the understanding of science, as previously stated, which never proceeds in regards to semantics, from theory to factual explanation. It leaves the impression that their conclusions are somewhat uncertain. At any rate, Descartes's second meditation is faulty; nevertheless, he can be accredited for producing the mathematical Cartesian coordinate system.

Baruch Spinoza (1632-1677 AD) pursued the truth despite the authority of the day. As a result, he found himself excommunicated from the Jewish establishment, and in this, one sees a typical case proving the obstacle truth is to religion. In the Will Durant book on page 130, he writes, "But, Spinoza had but one compelling desire–to reduce the intolerable chaos of the world to unity and order." So, did he succeed? In his book, *Improvement of the Intellect*, Spinoza wrote his reason for pursuing philosophy. "I determined at last to inquire whether there was anything which might be truly good, etc" (Durant, p. 130). But, this clarifies the intellectual problem for many philosophers including Spinoza. The word "good" refers to preference, not a truth about reality. There is no truly good or bad (except for the caveat of fairness as presented in chapter five of this book). The judgment of good is in the eyes of the beholder. For instance, when wars are won the winning side thinks that is good, while the losing side thinks that is bad. Consequently, Spinoza never realized the big picture of reality issues versus preference issues. Like his predecessors before him, he was extremely confused.

Immanuel Kant (1724-1804 AD) from Germany is known especially for his work on metaphysics and epistemology. His classic work, *The Critique of Pure Reason*, challenges the rationalists with their conclusion that knowledge is "a priori." This suggests knowledge is possible before sensory experience. Additionally, Kant argues against the empiricist view that all knowledge is only through sensory experience. To Kant, all knowledge is a combination. Let it be clear, the philosophers attempt to prove *a priori* knowledge with testing. But, the "scientific" testing reinforces my argument that the demarcations of philosophy issues versus science issues make no sense. Furthermore, Immanuel Kant was somewhat responsible for those faulty demarcations.

In regards to *a priori* knowledge, the fact that humans evolved from the animal kingdom supports a possibility of basic inherited fears and desires. But, when we observe the disagreements about reality among various people, it is clear *a priori* is at most very limited. This is because the facts required for real knowledge are empirically learned, not inherited. And many people get the facts wrong. Consequently, disagreements are inevitable. But the philosophy department needs *a priori* to support that they have achieved something, when, in fact, they are contributing to the confusion.

The German "philosopher" Arthur Schopenhauer (1788-1860 AD) is especially known for his essay called "The World as Will and Idea." It was initially rejected, says Will Durant, "because it attacked just those who could have given it publicity–the university teachers" (p. 233). This is an example proving that most institutions, including those of higher learning, are interested in truth only when it is perceived as self-serving. Ironically, this author too challenges the academic establishment.

In "The World as Will and Idea" Schopenhauer is searching for the essence of human motivation or the essence of man. This reality issue, not a preference issue, is currently studied by modern scientists (psychologists) as well. Consequently, is this philosophy? Furthermore, Schopenhauer concludes the unconscious will dictates to the intellect. I cannot disagree. But with more facts currently available, it becomes clear, from chapter three of this book, that the neo-cortex brain, where consciousness resides, is an intellectual tool produced by the processes of evolution (ultimately controlled by releases of pain and pleasure neurotransmitters), which serve the purpose of self-survival and species-survival statistically. Schopenhauer's "will" lacked the additional facts now available for a more complete explanation. Additionally, Schopenhauer suggests this "will" may be the metaphysical essence of all substances and forces as well. But this extrapolation is incorrect. The essence of all substances and forces are somewhat described by Einstein's energy and mass equivalence equation. Furthermore, as more facts are produced by physicists, a more detailed essence is likely to be produced.

Next in line, let us consider England's Herbert Spencer (1820-1903 AD). Spencer's thinking arose with the knowledge of biological evolution and Darwin's explanation of how that evolution could have occurred. Nevertheless, Spencer's *First Principles* Part One is entitled, *The Unknowable*, and suggests science will be ultimately limited to less than a complete knowledge. As per Spencer's words, "In its ultimate nature nothing can be known" reinforces the agnostic perspective. Perhaps this is correct. Nevertheless, because what we do know is from established facts and more facts are continually produced, the knowledge base continues to expand. Furthermore, enough facts have already been produced eliminating many theories as viable. For instance, the theory of an Almighty, Intelligent, Creator, God is thoroughly in contradiction to known facts as presented in chapters two and five of this book.

In regards to Spencer, an interesting definition of philosophy is expressed on page 276 of the Durant book. "The proper field and function of philosophy lies in the summation and unification of the results of science." As per Durant in regards to Spencer's perspective, Durant writes, "science is partially unified knowledge; philosophy is completely unified knowledge." From Spencer's perspective, science produces the facts, while philosophers put them together. This could almost be an intellectually acceptable definition of what philosophers do. But, there are problems with this because people called philosophers are often engaged in preference issues such as ethics and beauty, rather than big picture reality issues. Furthermore, many philosophers are, in fact, stating their own preferences rather than merely making an objective statement about reality. Please, consider Friedrich Nietzsche as an example.

Friedrich Nietzsche (1844-1900 AD) was born in Prussia and was appointed the chair of classical philosophy at the University of Basel after a brief, unsuccessful experience in the military. According to Durant, after discovering Schopenhauer's, "The World as Will and Idea," combined with his experience of the "magnificence" of the military, Nietzsche said, "I felt for the first time that the strongest and highest will to live does not find expression in a miserable struggle for existence, but in a will to war, a will to power, a will to overpower!" (p. 276) But this is a statement of preference, not an objective analysis of reality. After all, Nietzsche could have said, a will to overcome conflict, a will to peace. But instead, he preferred a will to war. And did this preference justify the First World War as well as Hitler's ambitions causing the Second World War with an uncountable victim tally? Ironically, according to Durant, "The sight of blood made him (Nietzsche) ill" (p. 305).

In regards to philosophers, Durant quotes Nietzsche, "The philosophers all pose as though their real opinions had been discovered through the self-evolving of a cold, pure, divinely indifferent dialectic: whereas, in fact, a prejudicial proposition, idea or 'suggestion,' which is generally their heart's desire abstracted and refined by them with arguments sought out after the event" (p. 316). In other words, many philosophers are intellectually justifying their own preferences rather than stating objective truths about reality, Nietzsche included.

Revolting against the clock-like mechanism of physics and materialism is Henri Bergson (1859-1941 AD) from France. Bergson argues against determinism and for free will with the following argument, according to Durant, "If determinists were right, and every act were automatic and mechanical resultant of preexistent forces, motive would flow into action with lubricated ease. But, on the contrary, choice is burdensome and effortful, etc. . . ." (p. 339). It may seem this way to Bergson, nevertheless, reality is understood by facts. And the related facts produced in physics cannot be ignored. At the small scale of physics, quantum theory implies there are inherent probabilities rather than direct causes and effects. But generally probability mathematics is employed when all the facts are not available to produce complete equations. Currently, this reality issue is unsettled because quantum theory is incomplete.

Currently, in regards to knowing the essence of reality, a "critical realism" founded by Karl Popper (1902-1994 AD) and (Roy Bhaskar 1944-) has iterated between modernism and postmodernism where modernism said reality can be known while postmodernism said reality can't be known. The Popper and Bhaskar conclusion is reality can be known to a certain, but not absolute degree. For what it's worth, this author agrees. From my perspective, reality is but a presentation of limited information. Nevertheless, within that information exposure, a certain truth can be separated from nonsense when comparisons are made. Thus, comparison is fundamental to any level of truth.

As previously mentioned in the first chapter of this book, the current philosophy that has come closest to making categorical sense of the big picture is the presentation of objective versus subjective and the presentation of empirical versus normative-values. But, as previously explained, objective often refers to the mind set of the observer rather than an acknowledgment of the object (reality), while subjective refers to both inaccurate perception and preference. Consequently, this is a failure to correctly separate out preference. In regards to the presentation of empirical versus normative-values, an empirical statement is one which can be potentially determined as true or false but fails to acknowledge reality while normative-values degenerate to the problematic terminology of subjective. None of this correctly accounts for the existence of reality issues versus preference issues. Consequently, the philosophy department has succeeded primarily at confusing everyone.

In review, some so-called philosophers have produced objective theories about reality (Thales's "all is water") while others unknowingly have stated their own preferences, (Aristotle's golden mean and Nietzsche's will to war). If this leaves you confused about what philosophy actually is, it should. But, what does Will Durant, the author of *The Story of Philosophy*, think philosophy is?

On page three of his book, he writes, "Specifically, philosophy means and includes five fields of study and discourse: logic, esthetics, ethics, politics, and metaphysics." Durant states logic to be "the study of ideal method in thought and research." But, how can there be two ideal methods of scientific versus philosophic? As previously presented, if their thinking is, certain issues are testable while others are not, then the thinking is faulty. From chapter one of this book, Einstein's "scientific" general relativity theory has had minimal testing while the "philosophic" theory of determinism has results from every duplicated test ever run. Therefore, from the Socratic Method of exposing contradictions, two distinct methods makes no sense as both cannot be ideal.

In regards to esthetics, according to Durant, "Esthetics is the study of ideal form, or beauty" (p. 3). But, an ideal form according to whom? There is no "ideal" here because esthetics is a preference issue where one person's beauty is another's ugly. Recently, I debated a PhD. philosopher in regards to this very issue. Her argument stated the testing of very young children proved there was a certain absolute in form preference. Though possibly true, this would indicate the process of evolution motivates us in certain preference directions; nevertheless, this argument fails to eliminate the esthetics issue as preference. This, because it does not matter if the reason for our preference is in our DNA or in our culture. Bear in mind, another direction of evolution could have altered our current preferences. Thus, beauty is a preference issue.

In regards to ethics, Durant states, "Ethics is the study of ideal conduct" (p. 3). But, here again the philosophers fail to realize this also is about preference. That said, with more than one person involved, we might consider "all sides are treated equally" as a reasoned criterion for preferences about conduct. But, otherwise, who are we to judge? That said, religions have justified their own moral preferences as absolute, based upon the failed All-mighty, Intelligent, Creator, God theory.

In regards to politics, according to Durant, "Politics is the study of ideal social organization" (p. 3). But the word ideal is a term of stated preference and politics is clearly about preference. Therefore, no ideal political system can exist. But again, if "all sides are treated equally" is the preference, then democracy is the resultant form of government. This happens because any dictatorship resulting from the aristocracy would inevitably benefit those ruling at the detriment of those ruled.

And finally, metaphysics, according to Durant, "Is the study of the ultimate reality of all things." In other words, what is the fundamental essence of reality? Clearly, this is an issue about reality as opposed to a preference issue. And basically this issue is resolved with Einstein's famous equation suggesting, all is energy.

Chapter Nine

What This World Is Really About!

So what is this world really about? It is about two issues, reality issues and preference issues. It is not, about philosophic and scientific methods incorrectly presented as issues by the academic institution. They have it wrong. But, the demarcations are self-serving. That is, they, the experts, are paid to study philosophic and scientific issues.

And how will they take on the challenge posed by this author? Will they say, where is his PhD? He is not one of us, experts. What does he know? It is ironic the academics have replaced the authority of a non-existent God (religion) with the authority of an academic degree. But the reader should know any legitimate counter argument should take on the specific merits of the argument.

One confused critic suggested this is just one of many approaches, as if there is something arbitrary about acknowledging the difference between reality and preference. To the contrary, is it an arbitrary approach to acknowledge the difference between the sun and the moon? Of course, the answer is no because there exists real differences. Likewise, an issue regarding reality is different from an issue regarding whether people like it or not. This acknowledgement is where the philosophers should have started. However, the philosophers got off to a wrong start when

Aristotle suggested there is a golden mean. When, in fact, the golden mean is just his stated preference. But, let it be understood that today's philosophers are equally confused. For instance, they think, based on certain test results, that beauty is an absolute rather than a preference. But why or how something is liked unanimously in a test does not alter it from preference.

So, why is it important to know there are reality issues and preference issues? Because, other than in fiction, all other issues falls into one of these two categories. As it turns out though, philosophers have had input to both categories without acknowledging so. Consequently, when someone says, "That's a philosophic issue," what do they mean? Is it unknowable? Is it an overview position? Is it untestable? Is it about reality? Is it about preference? Or is it some twisted combination? The truth is, the philosophers have successfully confused everyone, including themselves.

But if anything intelligent did come from a "philosopher," it is the Socratic Method of exposing argued inconsistencies. This automatically requires comparison of information. Therefore, an accurate or intelligent understanding of reality requires a comparison, not just a memorization of information. Please consider as fact chimpanzees have out performed humans in regards to certain memory tests. However, the human advantage for gaining intelligence is the neo-cortex frontal lobe, which allows for the comparison of information.

Let me make this perfectly clear. Intelligence, the understanding of fact-based reality, is a learned or acquired event. No one is born with intelligence. It must be learned, because, at the most fundamental level, the human intellectual brain is capable of only two operations. We can memorize information, and we can compare information. That is it. But, you might argue, what about imagination? My response is, imagination is a result of comparing information. Additionally, you might argue, what about intuition? My response is, intuition is a myth, the facts will not support it and the philosopher's *a priori* is at most very limited. Additionally, humans, as fact, possess no instincts or how-to knowledge as do the animals. Therefore, intelligence is primarily a learned event for the human species.

However, the educational system rewards students for memorizing data produced by their experts. Because, the IQ test is a statistical measurement of genetic memorizing capability, a high correlation exists between IQ test scores and academic scores. But, for the most part, the educational system has left graduates unknowingly confused while dependent upon their experts. As proof of the confusion, many of their highest IQ PhD graduates go to church on Sunday and believe what they hear.

Nevertheless, religion is an insult to intelligence. That is, religion is not a test of faith; it is a test of intelligence. At least, any religion that suggests there is an All-mighty, Intelligent, Creator God. The facts conceded in the book, *Darwin on Trial*, by the Christian author Phillip Johnson, as presented in chapter two of this book, are in dire contradiction to a theory suggesting the existence of an All-mighty God. Furthermore, the facts conceded are in dire contradiction to a theory suggesting the existence of an intelligent God. And lastly, the facts conceded are in dire contradiction to a theory suggesting the existence of a Creator God. Whether the promoters of religion concede or not, their belief is a theory about reality where a theory must explain the related facts. But the facts get in the way of their theory. And that explains why religion is promoted as a faith issue. But how else can something that does not exist be promoted?

However, the "intellectual" agnostics are unsure. But I wonder if the agnostics are also unsure of the existence of Santa Claus. And maybe they believe the moon is actually made of green cheese. Or perhaps we are in the matrix, as in the Hollywood produced movie. It is all too confusing. The agnostics just don't know. Or perhaps the educational system has successfully left many confused when so many of their graduates claim to be agnostic or religious.

Nevertheless, the conservatives use religion in order to control their vast human herd. From every church on Sunday, the brainwashed followers are primed to align with the conservative preference, making Rush Limbaugh's sales pitch an easy one. But, what's especially amusing is when a financially poor social conservative votes for a fiscal conservative. Do they not realize the fiscal conservative will impose self-serving, trickle-down, failed economic theory, which benefits those born into wealth, while being contrary to a poor person's self-interest? Consequently, the former Reagan administration is held up as a model for modern conservatism in America.

But for all the current Tea Party chatter about the gross national debt, the hard facts support the steep debt incline first occurred during the Reagan administration. From 1950 to 1981, the federal debt curve was basically level. But, during the eight-year Reagan administration the federal debt started at $994 billion and ended at $2,867 billion. This was an increase of nearly three hundred percent. Nevertheless, Reagan's trickle-down economic policy was advantageous to rich campaign donors unconcerned about the nation's debt. Consequently, both conservative George H. Bush and his son George W. Bush continued deficit spending while only during the liberal Clinton administration did this trend reverse.

But now in 2014, we have a black Democrat in office. The conservatives need a reason to replace him with one of their own. Given the choice between reducing the nation's debt or maintaining some of the lowest tax rates for the wealthy in

American history, the Republicans opted for the latter, proving they have no actual concern about the debt level. It was just a ploy to hammer the current Democratic administration. Bear in mind, Obama's eight-year administration is on track to double the national debt. But, Reagan's eight-year administration tripled it!

In regards to social conservatism, there are many sins, with hell to be paid. But, heaven and hell are just man-invented extensions of nature's pleasure and pain neurotransmitters, which cause our fundamental motivations. The intellectual psychologists and philosophers can't seem to figure this out, but those promoting religion did so thousands of years ago. Therefore, man-invented heaven and hell are the ultimate psychological carrot-and-stick approach to controlling the human herd.

With such deception practiced, it is quite ironic they hold themselves up as the "moral" majority. Furthermore, when I hear a conservative politician talk about his or her principles it's laughable. What principles? Certainly the truth, that which is consistent with the available facts or fairness, all sides are treated equally, has nothing to do with their stated principles. If the truth is spoken, all politicians are but servants to their campaign donors. Their stated principles usually are just cheap and meaningless sales pitches.

In regards to the principle of fairness, please consider the inequity of the tax code. For those born into wealth, the federal estate tax rate (moved up during the George W. Bush administration) allows for five million dollars tax free from a single individual to each recipient. Ten million dollars is allowed tax free from a married couple to each recipient. Bear in mind, the recipient produces no goods and services for this free money. And where no goods and services are produced, maximum inflation occurs because the purchasing power (demand) to supply ratio is at a maximum similar to printing extra money. Because, we live within an economic, interconnected system all others feel the impact. Now consider a person born into poverty. Any money they get they must earn at rates that punish them for producing. And of course, the conservatives fight any legislation that might increase minimum wage while they promote legislation that produces tax loopholes benefiting the rich. Is it any wonder the rich get richer?

But the conservatives have their canned response in regards to inheritance, this money has already been taxed by the earner. Therefore, any additional taxation would qualify as double taxation. However, person two, the recipient is not person one. Consequently, it's not double taxation as they suggest. However, their argument logically reduces to the idea that money should not be taxed when given away. However, to be consistent, let everyone give money in exchange for gifts of goods and services. Consequently, no taxes would ever be paid with failure of government assured.

Nevertheless, if the preference for a modicum of fairness were in play, the first ten million dollars earned by each non-inheritor would be federal tax free, also. Of course this will never happen because all those campaign contributions by the rich creating self-serving laws for the rich. Bear in mind, if everyone has ten million dollars, then no one has ten million dollars. Purchasing power is relative, and the rich conservatives know this.

But you will never hear this on the Rush Limbaugh conservative radio talk show. Bear in mind, Rush Limbaugh never had to fight it out at the bottom. His dad was, coincidentally, half owner in a significant radio station. Likewise, the right-leaning cable Fox News Network is owned by Rupert Murdoch, who inherited an entire newspaper organization from his dad. Consequently, their right-wing sales pitch benefits themselves and a relatively few others born into wealth and status. But the other 98 percent or so of the population must somehow be controlled by the practice of deception.

At this point, a critic might argue, all this is "just your opinion." But what does this nebulous "opinion" actually mean? Does opinion mean a stated preference? Or does it mean this is just a subjective perception? Or does opinion mean only the experts know with certainty? To the contrary, this entire book is an analysis of reality based upon the available and directly related facts. That said, anyone could be wrong, including me. But if someone wants to argue, reality is not understood according to facts, I look forward to that debate. And when all the facts are considered, the two issues of reality and preference emerge.

Bibliography

Berra, Dr. Tim. *Evolution and the Myth of Creationism*, CA: Stanford University Press, 1990, pp. 2, 4.

Durant, Will. *The Story of Philosophy*, NY: Simon and Schuster, 1961, pp. 3, 7 106, 130, 233, 276, 305, 316, 339.

Ferris, Timothy. *The Whole Shebang*, NY: Simon and Schuster, 1997, p. 32.

Green, Brian *The Hidden Reality*, page 9, NY: Vintage Books/ Random House, 2011, p. 9.

Johnson, Phillip E. *Darwin on Trial*, second ed., IL: Inter Varsity Press, 1993, pp. 16, 34, 49, 52, 54, 81, 82.

Limbaugh, Rush. *The Way Things Ought To Be*, NY: Pocket Books/ Simon and Schuster, 1992, p. 2.

Review Requested:

If you loved this book, would you please provide a review at Amazon.com?